Be a Plant SCIENTIST

• Question • Experiment • Discover

By Alix Wood

Ruby Tuesday Books

Published in 2024 by Ruby Tuesday Books Ltd.

Copyright © 2024 Ruby Tuesday Books Ltd.

All rights reserved. No part of this publication may be reproduced in whole or in part, stored in any retrieval system, or transmitted in any form or by any means, electronic, mechanical, photocopying, recording, or otherwise, without written permission from the publisher.

Editors: Ruth Owen & Mark J. Sachner
Design: Alix Wood
Production: John Lingham

Photo credits:
AdobeStock: 6C, 6B, 19TR; Alamy: 4TR (Pat Canova), 10BR (ImageBroker.com), 18B (Auscape International Pty), 20R (Tony Watson), 25T (Susie McCaffrey); Dreamstime: 30R (Oleh Malshakov), 31R (Photographyfirm); Ruby Tuesday Books: 7, 13, 14, 17B, 21, 31L; Shutterstock: Cover TL (enjoy photo), Cover TC (Erika Kirkpatrick), Cover TR (Smeerjewegproducties), Cover BL (Mohamed Fox), Cover BC (Barbara Ash), Cover BR (Triff), 1 (Lebid Volodymyr), 3 (xpixel), 4TL (ozgurshots), 4B (ifong), 5TL (PERO Studio/Nitr), 5TC (Esin Deniz/Resource Image), 5TR (Body Stock/Creative Photo Shop), 5B (Pixel-Shot/gresei/chictype Montreal/Spalnic/Photoongraphy/Khvost), 6T (snapgalleria), 8T (Lebid Volodymyr), 8BL (Aldona Griskeviciene), 8BR (giedre vaitekune), 9TL (Bigc Studio), 9TR (Katerina Maksymenko), 9B (Gargonia/AjayTvm/Ihor Hvozdetskyi/Erkki Makkonen/Paul Maguire), 10T (xpixel), 10BL (Mariyana M), 12T, 12BL (Elena D.), 12BR (ShantiawanPb), 15, 16T (Smit), 16TR (LiuSol), 16B (NeoTunesPhoto), 17T (Leptospira), 18T (kram-9), 19, 20L (Frank11), 22–23, 24T (Lou WOZ), 24B (Gordon Magee), 25B, 26TR (SGr), 26B (TinnaPong), 27, 30L (Tobia Tropper/Triff/Vibe Images), 30R (Bess Hamitii/Oleksandrum), 31L (morozv/Oleksandr Lytvynenko/Studio 888), 31R (Pressmaster), 32 (Gail Johnson); Superstock: 10BL (Science Photo Library); Alix Wood: 11.

ISBN 978-1-78856-434-2

Printed in Poland by L&C Printing Group

www.rubytuesdaybooks.com

Contents

Let's Investigate Plants 4
Roots in Action 6
Super Stems 8
Stems in Action 10
Let's Look at Leaves 12
Did You Know That Plants Can Cook? .. 14
Falling Leaves 16
Soil Buffet 18
Flowers Need Busy Bees! 20
Super Seeds, Fantastic Fruits 22
How Do Seeds Find a Home? 24
Room to Grow 26
Let's Talk Plants 28
Glossary 30
Index 32

Let's Investigate Plants

What do a tiny daisy and a huge oak tree have in common? They are both plants!

Daisy plant

Oak tree

Plants are living things that can use sunlight to make their own food.

Most plants have **roots** that grow underground in **soil**.

They also have **stems**, leaves and flowers that produce **seeds**.

Flower, Leaf, Stem, Soil, Roots

Plants are very useful to people.
We use them for food, and to make fabric and furniture.

Lettuce plants

Cotton plants

Pine trees

Salad

Cotton T-shirt

Wooden table

Which of these things were made from plants?

Can you find something that was made from a plant in your home or classroom?

To find answers and more information, turn to page 28.

Now, it's time to investigate the amazing lives of plants and . . .

. . . be a **Plant Scientist**.

Roots in Action

Most plants need soil, sunlight, air, water and **nutrients** to live and grow.

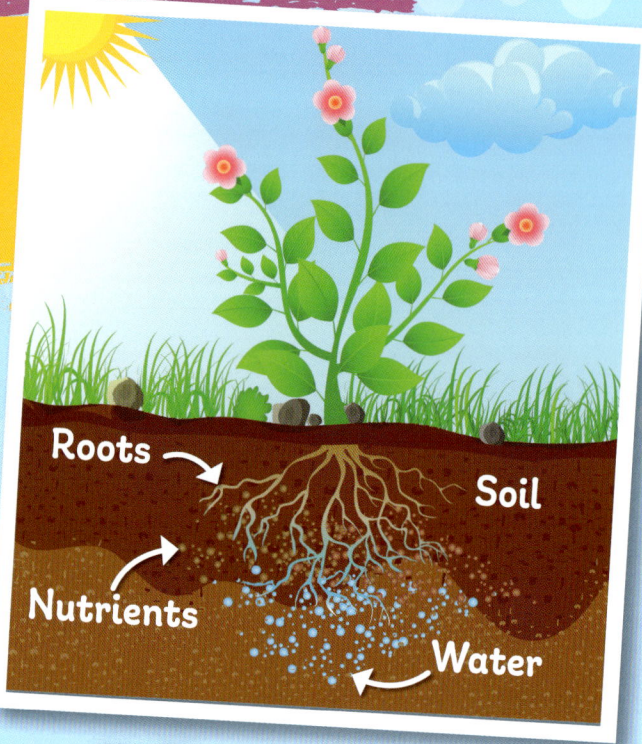

How do plants get the water they need? They use their roots.

A plant's roots suck up water and nutrients from the soil.

Some plants have roots that grow out in all directions.

A plant's roots hold the plant in the soil and stop it falling over.

Other plants, such as carrot plants, have a long, thick taproot.

We eat a carrot's orange taproot!

Let's see how roots work.

Make a model of a tree and see how water flows from its roots all the way up to its leaves.

You will need:
- 1 large sheet of strong kitchen towel
- Scissors
- A ruler
- A green marker pen
- A toilet roll tube
- A small glass or jar
- Double-sided tape
- Water

1) Begin by cutting narrow strips, about 7.5 cm long, in one side of the paper.

2) Twist each narrow strip. These are your tree's roots.

3) Now cut five wider strips on the opposite side of the paper. These strips are your tree's branches.

4) Draw leaves at the end of each branch. Twist the branches.

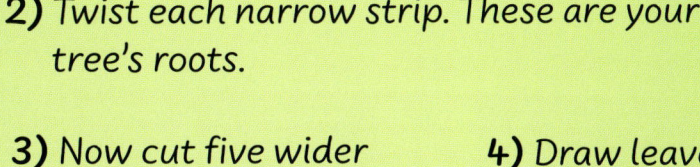

5) Roll up the paper and slide it into a toilet roll tube. The tube is your tree's protective **bark**.

6) Stand your tree in a glass or jar. Tape it in place if needed. Carefully pour 2.5 cm of water into the container. Keep checking your tree for the next hour.

What do you observe happening?

To find answers and more information, turn to page 28.

How is this similar to what happens in a real plant?

7

Super Stems

Most plants have stems that grow out of the soil. Trees have a thick main stem called a trunk.

Trunk

A tree's woody trunk is covered with bark. Bark protects the tree from hot sun, rain, snow and munching animals.

Some plants have thin or weak stems.

They climb up walls, long sticks or other plants to reach sunlight.

Some climbing plants use tendrils to hold on.

Tendril

Climbing plant

Herbs are plants with thin, soft stems.

Basil

Shrubs are plants with several woody stems.

Rose shrub

Pumpkin plant

Some plants, such as pumpkins, have soft, weak stems that creep along the ground.

Let's go on a plant scavenger hunt.

A good scientist is always observing the world around them. Visit a garden, park or forest and look carefully at the plants.

Can you find these items? You get **10 points** for every one you spot!

10 Rough tree bark

10 A tendril

10 A shrub with woody stems

10 A creeping plant

10 A climbing plant

Stems in Action

The job of a plant's stems is to carry water and nutrients to the plant's leaves, flowers and fruits.

Leaf

Twig

Apple tree branch

Flower

A tree has stems called branches that grow from its trunk. Thinner stems called twigs grow from the branches.

Water and nutrients flow up from a plant's roots.

Then they flow along thin tubes in a plant's stems.

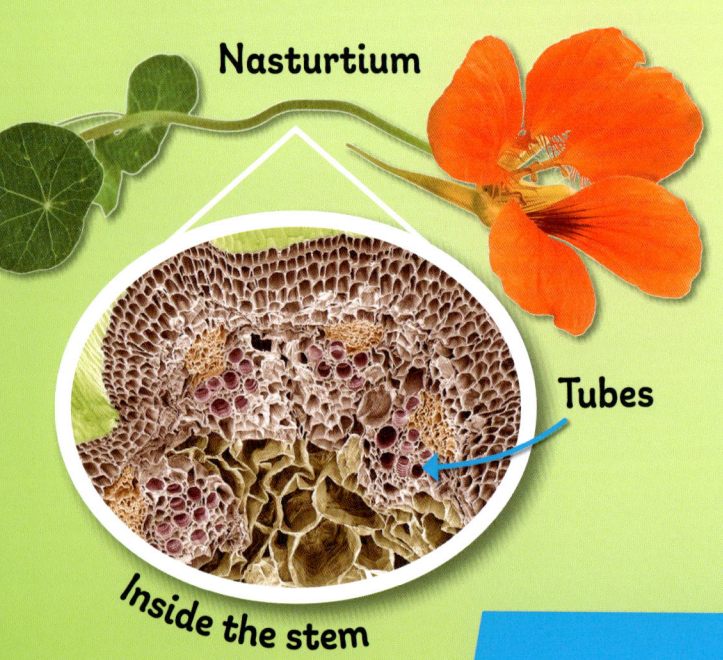

Nasturtium

Tubes

Inside the stem

A giant bamboo plant's stems grow almost 1 metre in a day!

10

Can a bean plant solve a maze?

Plants grow up towards the sunlight they need to survive.

Build this maze, and see if a bean **seedling** can grow its way out to the light.

You will need:
- A bean plant growing in a flowerpot
- A shoebox
- Scissors
- Some card
- Tape
- A small watering can

1) Remove the lid from the shoebox and stand the box on end. Cut a hole in the top of the box.

2) Cut two rectangles of card that are the same size as the end of the shoebox.

3) Fold each piece of card under (as shown). Then tape the pieces inside the box, one on each side.

4) Place your bean seedling at the bottom of the box.

5) Replace the lid. Stand the box on a windowsill. Water your seedling every other day to keep the soil moist.

Take photos or draw your bean plant as it grows.

Did the seedling solve the maze and grow out of the box?

To find answers and more information, turn to page 28.

11

Let's Look at Leaves

Leaves can be round, oval, pointed, jagged and even heart-shaped. Some are smooth. Others are waxy or hairy.

Leaves may look very different, but they all have one thing that's the same — **veins**.

Veins are thin tubes that carry water from the stem into the leaf.

Veins

Stem

Some leaves have veins that look like a net pattern.

Others have veins that run in straight lines.

How do leaf veins work? Let's investigate!

You will need:
- A small glass
- Water
- Food colouring
- A spoon
- Scissors
- Leaves cut from a plant (with permission) — pale leaves or ones with white veins work best
- Paper and wax crayons

1) Half fill the glass with water. Add 3 drops of food colouring and mix it with a spoon.

2) Choose a leaf and take a photo of its veins.

3) Stand the leaf in the water, stem first.

4) If you wish, repeat steps 2 and 3 with other leaves.

5) Check your leaves every day for three days. Look closely at the veins and compare them to your photos.

What do you observe? Describe what is happening to the leaves.

Make a Leaf Rubbing

Place a leaf on a flat surface so its underside and veins are showing.

Press a sheet of printer paper over the leaf.

Rub a wax crayon over the leaf shape to reveal the pattern of the veins.

To find answers and more information, turn to page 28.

Did You Know That Plants Can Cook?

Plants use water and sunlight to make their own sugary food called glucose.

A Plant's Ingredients for Cooking

A plant's leaves contain green stuff called **chlorophyll**.

It's chlorophyll that makes plants a green colour.

Chlorophyll traps the energy in sunlight.

The leaves take in **carbon dioxide gas** from the air through tiny holes that open and close.

Water flows from the roots up into the leaves.

The leaves use the energy from sunlight to turn water and carbon dioxide into food.

Do plants really need sunlight to survive?

What will happen if we keep a plant in a dark place? Let's investigate! To do this experiment you will need two plants of the same size and type.

You will need:
- 2 identical potted plants (herb plants from a supermarket work well)
- A sunny windowsill
- A dark cupboard
- A small watering can

1) Place one plant on a sunny windowsill. Place the other plant in a dark cupboard.

2) Give each plant a little water every day to keep the soil in the pot moist. You must give each plant exactly the same amount of water.

3) After two weeks, take the plant from the cupboard. Compare your two plants.

How are the plants still alike? How are they different?

To find answers and more information, turn to page 29.

The tiny holes in leaves are called stomata.

 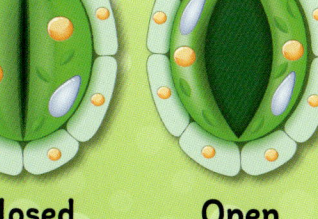

Closed   Open

As a plant makes its food, it also makes a gas called oxygen. It releases the oxygen from its stomata.

Have you heard of oxygen? Why is it important?

Falling Leaves

Some trees and shrubs lose their leaves in autumn. Why?

In winter there are fewer hours of daylight.
Often there is less rain and the soil freezes.

Without sunlight and water, a tree's leaves can't make food.
The leaves start to change colour, and fall to the ground.

Trees that drop their leaves in autumn are called **deciduous** trees. They save energy all winter, and then grow new leaves in spring.

When a tree stops making food, it also stops making green chlorophyll.

Leaf with chlorophyll

Dead leaf

Then it's possible to see a leaf's other colours that are normally hidden by green.

What happens once a leaf falls from a tree?

You will need:
- 4 freshly picked green tree leaves
- Baking paper
- Scissors
- Some heavy books

1) Pick four green leaves from a tree.

2) Open a heavy book. Cut a piece of baking paper that covers the pages.

3) Place two leaves on the right-hand side of the paper. Close the book. Place more books on top.

4) Put the other two leaves somewhere safe where air can reach them.

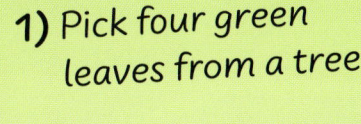

What do you predict will happen to the two sets of leaves?

5) After one week, check all four leaves and compare them.

What do you observe happening to the leaves?

How are the two sets of leaves the same or different?

6) Put the leaves back in their places. Wait for one week, and check them again.

To find answers and more information, turn to page 29.

Soil Buffet

Plants grow in soil, and take in nutrients from soil with their roots. But what is soil made of?

Soil is made of tiny pieces of rock.

It also contains the **rotting** remains of dead plants and animals.

Rock →

Twig ↘

Dead leaf ↙

Worm hole

Worm

Roots

Air and rainwater collect in holes and tunnels made by worms.

All the rotting matter in soil contains nutrients that feed plants.

18

Be a Nutrient Detective

What happens if a plant does not have enough nutrients?

You will need:
- 2 flowerpots
- Some garden soil
- Potting compost
- A small spade
- 4 bean seeds
- A watering can

1) Fill a flowerpot with soil from a dry or dusty spot in a garden or your school playground.

Garden soil

2) Fill a second flowerpot with potting compost.

Examine the two types of soil. How are they different? How do they feel if you squish them in your hand?

Potting compost

3) Push two bean seeds into each pot — about 2.5 cm deep.

4) Water the two pots and place them in a warm, sunny spot.

5) Keep the soil moist in both pots. Make sure you give each pot exactly the same amount of water. Observe your plants for up to four weeks.

Garden soil   Potting compost

Which plant looks the most healthy?

Do you see any yellow leaves? What could this mean?

To find answers and more information, turn to page 29.

Flowers Need Busy Bees!

Flowers have an important job to do — they make seeds! To do this, a flower must be **pollinated**.

Dusty pollen from a flower's anthers must land on a flower's stigma.

Then the flower is pollinated.

Anther

Stigma

Cherry tree flower

Bee

Pollen

Insects, such as bees and butterflies, carry pollen from flower to flower on their bodies.

However, a flower can only be pollinated with pollen from a flower of the same type.

Insects visit flowers to feed on pollen and a sweet liquid called nectar.

Be a Cotton Ball Bee

Can you do a bee's job and help pollinate flowers?

You will need:
- Paper
- A black marker pen
- Scissors
- Coloured chalks
- A cardboard box
- A towel
- A cotton wool ball taped to a skewer (to be a bee)

1) Draw six large flower shapes and cut them out.

2) Using chalk, colour the centre of two flowers in blue, two in red and two in yellow. Colour them really thickly, so they are very dusty.

3) Place a flower of each colour on a table.

4) Place the box on its side and drape a towel over it. Put the other three flowers inside the box. Without peeking, shuffle the flowers.

5) Dab the cotton wool ball bee in the centre of a flower on the table.

6) Now put your hand in the box and choose a flower to pollinate. Don't peek! Dab the bee on that flower. Check the bee.

Remember! A flower needs pollen from a flower of the same kind to be pollinated.

Did you pollinate the right flower? If not, try again, Bee.

Super Seeds, Fantastic Fruits

Once a flower has been pollinated, it produces seeds. The seeds form inside a protective fruit.

Tomato plant flower

Tomato plant fruits

Some foods that people call vegetables are actually fruits.

That's because they contain seeds.

True vegetables are the stems, leaves or roots of plants.

Tomatoes, cucumbers and squashes are all fruits.

Some plants grow a fruit with just one seed inside. Others grow fruits with many seeds.

Plum seed

Kiwi fruit seeds

Let's Investigate Seeds!

Not all fruits are juicy like a plum.

A pea pod is the fruit of a pea plant. Inside the pod are peas, which are seeds.

Pea pod

Ask an adult to help you cut open an apple and a melon. Then count the seeds.

How many seeds do you estimate will be in each fruit?

A melon has many seeds. Can you think of a way to make the counting easier?

Cut open a strawberry. How many seeds can you see inside? What do you observe about this fruit?

Fruit or Vegetable?

Sort these pictures into whether they are of a fruit or a vegetable.

Pumpkin

Beans

Peppers

Celery

Avocado

Broccoli

To find answers and more information, turn to page 29.

How Do Seeds Find a Home?

A seed needs a place to grow so it can become a new plant.

Seeds

Some seeds **disperse**, or spread, by hooking on to an animal's fur.

When the seeds fall off, they are in new growing places away from their parent plant.

When a bird eats berries, the seeds inside drop to the ground in its poo.

Nutrients in a bird's poo can even help feed a seedling as it grows.

Some seeds grow in pods that explode! The seeds are blasted to their new homes.

Poppy flower

BANG!

Bud

Dried seedpod

Exploding Balloon Seedpod

How far will the seeds travel?

1) Begin by stretching the balloon to loosen and soften it. Next pull the balloon's opening over the neck of the funnel.

2) Pour the seeds into the funnel. Shake the funnel to make the seeds fall down into the balloon.

3) Ask your adult helper to blow up the balloon. Tie up the opening.

4) Stand outside and burst the balloon with the pencil.

5) Examine the area to discover how far your seeds have dispersed. Measure the distances.

You will need:
- A balloon
- A funnel
- 1 cup of wild bird seeds
- An adult helper
- A pencil
- A tape measure

Make sure you collect up all the pieces of burst balloon.

Did any of your seeds land in a good growing place?

How far away did the furthest seed land?

Room to Grow

Why do seeds need to move away from their parent plants? Because seedlings need room to grow.

Seedlings

If lots of seedlings grow too close together, they are in a competition.

They compete for their share of water and nutrients.

Some plants get too tall when they grow close together.

Their stems become too long and weak as they compete for light.

Plant diseases can easily spread when plants are too close together.

Long, thin stems

Do more seeds mean more plants?

What will happen if you plant lots of radish seeds close together? Let's investigate with this experiment.

You will need:
- 3 plant pots, about 10 cm wide
- A marker pen
- Potting compost
- A packet of radish seeds
- A notebook and pencil
- A watering can

1) Label the three pots 3, 10 and 30. Fill them with soil.

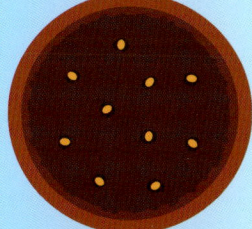

2) In the first pot (number 3), make three finger holes 1 cm deep. Plant a seed in each hole.

3) Repeat step 2 in the second pot, but this time plant 10 seeds. Finally, plant 30 seeds in the third pot.

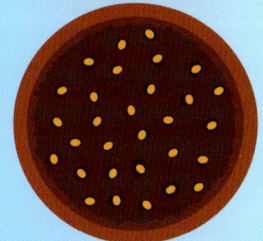

4) Sprinkle soil over the three pots to fill the holes.

5) Keep the pots in a warm, sunny place. Water them to keep the soil moist.

Which pot do you predict will grow the strongest radishes?

Write down your prediction.

6) After four weeks, pull the plants from the soil. Put the radish plants next to their pots.

Which plants have the healthiest leaves?

Do your results match your prediction?

Which plants have the biggest red radishes?

What do your results tell you about plant competition?

Let's Talk Plants

Did you enjoy being a plant scientist? Let's check out some answers and discover more cool things about plants.

Page 5:
All of the things were made from plants!
Ketchup is made from tomatoes, which are the fruits of tomato plants. The fries are made from potatoes, which are underground parts of a potato plant. The outside of a pencil is made of wood from trees. The pink eraser is made of rubber, which is a thick liquid made by rubber trees. Orange juice comes from oranges that grow on trees. The pyjamas are made of cotton from a cotton plant. The cookie is made using flour (from wheat) and sugar that both come from plants. The cookie's chocolate filling is made from the seeds of a cacao tree. The guitar's main body, the long neck and other parts are made of wood.

Page 7:
Did the water slowly flow up from the model's roots to the branches and leaves? In a real plant the roots suck up water. Then the water flows through tiny tubes in the stems and up into the plant's leaves.

Page 13
Did the veins in your leaves change colour? That's because the leaf was taking in water from the cup with its stem — just as it takes in water from a twig or branch.

Could you see tiny veins once they were coloured?

Page 11:
Did your bean seedling twist and turn around the pieces of cardboard to reach the hole? A plant needs sunlight to make its food (see page 14), so plants always try to grow towards light.

Page 15:
Your plant in the dark cupboard probably looks less healthy than the one on the sunny windowsill. Are the plant's leaves turning yellow? Are its stems floppy and dying? That's because the plant has no sunlight and cannot make food for energy.

Plants also make oxygen, which is a gas in the air that humans and animals need to breathe.

A plant's ability to make its own food using sunlight is called photosynthesis.

Page 17:
The leaves left in the air will start to dry out. Over time, they will curl up and became brown and crumbly as they rot and break down. This is what happens when leaves fall to the ground. Then they become part of the soil.

Decomposing leaf

Pressed leaf

The leaves pressed between the book's pages were protected from the air. This kept them from getting dry and crumbly so quickly. They probably kept some of their green colour, too.

Page 19:
The two pots of seeds had the same amount of sunlight and water. Did they grow in a different way or look different? If so, the only difference was the soil.

The dry or dusty soil from a garden or playground did not have as many nutrients as the fresh potting compost. This made it harder for the plants in that pot to grow and be healthy. Did their leaves turn yellow? That was because they needed more nutrients.

Page 23:
A watermelon may have up to 800 seeds! To make counting easier, you could ask friends to help you. You could make a heap of 20 seeds and then make similar size heaps, and count the heaps. Or you can simply estimate, which is making a good guess.

Each tiny yellow part on the outside of a strawberry contains a seed. There can be about 200 seeds on a strawberry.

Fruits: *pumpkins, peppers, avocados and green beans (the seeds — or beans — are inside the pods)*
Vegetables: *broccoli, celery*

Glossary

bark
A protective covering that grows on a tree's trunk, branches and twigs.

carbon dioxide gas
An invisible gas in the air that plants use to make food. When humans and animals breathe out, they release carbon dioxide into the air.

chlorophyll
The substance in leaves that traps sunlight. Chlorophyll gives plants their green colour.

deciduous
Trees that drop their leaves in the fall and grow new ones in spring. Apple trees and oak trees are deciduous.

disperse
To spread out. Some seeds disperse from their parent plant on the wind.

nutrient
A substance that a living thing needs to grow and be healthy. Plants get nutrients from soil.

pollinate
To carry pollen from flower to flower. Insects, birds, bats and other animals help pollinate plants. The wind carries pollen, too.

roots
Plant parts that are usually under the ground in soil. Roots take in water and nutrients. Roots also hold a plant in the soil.

rotting
Breaking down and becoming mouldy. Rotting leaves, flowers and fruits become part of the soil.

seed
A tiny part of a plant that contains all the material needed to grow a new plant.

seedling
A new, young plant that sprouts from a seed.

soil
Black or brown crumbly material that most plants need to grow in. Soil is made of tiny pieces of rock and rotted material such as dead plants and animal poo.

stem
The main part of a plant's framework that grows out of the ground, or a pot of soil. Thinner stems usually grow from a main stem.

vein
A thin tube inside a living thing. Water and nutrients flow through a plant's veins. Blood flows through a person or animal's veins.

31

Index

C
carbon dioxide gas 14
chlorophyll 14, 17

F
flowers 4, 10, 20–21, 22, 25
fruits 10, 22–23, 28–29

L
leaves 4, 7, 10, 12–13, 14–15, 16–17, 18–19, 22, 27, 28–29

N
nutrients 6, 10, 18–19, 24, 26, 29

P
pollination 20–21, 22

R
roots 4, 6–7, 10, 14, 18, 22, 28

S
seedlings 11, 24, 26, 28
seeds 4, 20, 22–23, 24–25, 26–27, 28–29
soil 4, 6, 8, 16, 18–19, 29
stems 4, 8–9, 10, 12, 22, 26, 28–29
sunlight 4, 6, 8, 11, 14–15, 16, 26, 28–29

T
trees 4–5, 7, 8–9, 10, 16–17, 28

V
veins 12–13, 28